HRW

ALGEBRA ONE

INTERACTIONS

COURSE 1

TECH PREP RESOURCES

HOLT, RINEHART AND WINSTON
Harcourt Brace & Company

Austin • New York • Orlando • Atlanta • San Francisco • Boston • Dallas • Toronto • London

Representing Patterns With Tables and Graphs

AirTech Aircraft Project

Project Description

The goal of this project is to have students solve constant speed problems, organize their results, and communicate them in an easily interpreted format within a predetermined time period.

Life/Work Skills

- Identifying an appropriate problem-solving strategy
- Distinguishing between meaningful and extraneous information
- Communicating effectively by using charts and graphs
- Working as part of a team

Introducing the Project

You may want to review some of the common ways of visually presenting information: bar graphs, circle graphs, line graphs, and tables. Prepare a few of your own charts as samples of high quality work. Ask groups to examine your samples and to answer the question, What makes an effective chart or graph?

Time Line *3–5 days*

Select a definite number of days that students can spend on their projects to ensure that they manage their time. Students should spend their first day seeking an understanding of the problem and defining the subskills required to create a finished product. Students may have initial difficulty in understanding how to approach the task. Encourage groups to act out each scenario as a means of identifying the salient information for each of the different scenarios. Make sure that each group completes a Daily Planning Log to ensure conscious allocation of its resources of time and manpower, as well as to provide you with an assessment of each day's work. Students should use about half of their available time to create graphs and tables for their presentation.

Answers

Answers may vary. A line graph shows how the flight range increases with altitude. Interpolation must be used for altitudes of 3000 and 5000 feet. A sample table with flight ranges in nautical miles is shown.

Altitude (ft)	Headwind (knots)			No wind		Tailwind (knots)		
	30	20	10	with reserve	no reserve	10	20	30
2000	513	555	597	524	639	680	722	764
3000	520	561	603	529	645	687	728	770
5000	530	572	614	538	655	697	739	781
8000	547	589	630	551	672	714	755	797
12,000	586	630	674	596	718	762	806	850

Using the "Pythagorean" Right-Triangle Theorem

Straiter's Stairs Project

Project Description

The goal of this project is to encourage students to use their creative-thinking skills to generate a model within the necessary specifications.

Life/Work Skills

- Identifying an appropriate problem-solving strategy
- Applying creative-thinking skills
- Understanding complex relationships
- Acquiring necessary information
- Working as part of a team

Introducing the Project

You may want to review how the "Pythagorean" Right-Triangle Theorem can be used to determine the length of any one of the sides of a right triangle given the lengths of the other two sides. Prepare a few of your own scale drawings of right triangles with the actual lengths of all three sides clearly labeled. Ask groups to examine your drawings and to answer the question, What makes these drawings easy to read and clear?

Time Line *3–5 days*

Select a definite number of days that students can spend on their projects to ensure that they manage their time. Students should spend their first day identifying possibilities for the size specificaitons and necessary formulas. Encourage groups to brainstorm to discover the information they need and to visualize a model either in their minds or as a draft drawing. Make sure that each group completes a Daily Planning Log to ensure conscious allocation of its resources of time and manpower, as well as to provide you with an assessment of each day's work. Students should use about half of their available time to identify the data that they will use and to create the final templates.

Answers

Answers may vary. A sample template for a total stair height of 3 feet is shown with data labels.

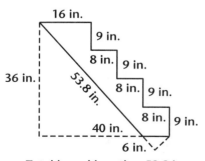

Total board length = 59.8 in.

Solving Proportion and Percent Problems

Growth Center Project

Project Description

The goal of this project is to encourage students to use their skills to solve proportion and percent problems (including various modified conditions) and to use their solutions to create a chart of prices for seed mixtures.

Life/Work Skills

- Identifying an appropriate problem-solving strategy
- Distinguishing between meaningful and extraneous information
- Communicating effectively by using charts and graphs
- Working as part of a team

Introducing the Project

You may want to review how to solve proportion and percent problems and how to organize the results into an easy-to-read chart. Prepare a few of your own charts as samples of high quality work. Ask groups to examine your samples and to answer the question, What makes an effective chart?

Time Line *3–5 days*

Select a definite number of days that students can spend on their projects to ensure that they manage their time. Students should spend their first day seeking an understanding of the problem and defining the subskills required to create the finished product. Encourage groups to act out each scenario as a means of identifying the important information. Make sure that each group completes a Daily Planning Log to ensure conscious allocation of its resources of time and manpower, as well as to provide you with an assessment of each day's work. Students should use about half of their available time to create charts for their presentation.

Answers

Answers may vary. A sample chart is shown.

	5-pound bag	12-pound bag	25-pound bag	Bulk (per pound)
Economy Sun Mixture	$3.90	$8.89	$18.14	$0.71
Economy Shade Mixture	$3.43	$7.83	$16.30	$0.62
Standard Sun Mixture	$4.16	$9.48	$19.40	$0.76
Standard Shade Mixture	$3.74	$8.54	$17.40	$0.68
Deluxe Sun Mixture	$5.20	$11.86	$24.18	$0.95
Deluxe Shade Mixture	$3.95	$9.02	$18.38	$0.72

Representing Linear Functions by Graphs

Mobile Communications Project

Project Description

The goal of this project is to encourage students to identify linear relationships and to generate graphs of linear functions that effectively communicate information in an easy-to-read format.

Life/Work Skills

- Identifying an appropriate problem-solving strategy
- Distinguishing between meaningful and extraneous information
- Communicating effectively by using charts and graphs
- Working as part of a team

Introducing the Project

You may want to review how to identify linear relationships and methods of graphing linear functions. Prepare a few of your own graphs as samples of high quality work. Ask groups to examine your samples and to answer the question, What makes an effective graph?

Time Line *3–5 days*

Select a definite number of days that students can spend on their projects to ensure that they manage their time. Students should spend their first day seeking an understanding of the problem and defining the subskills required to create the finished product. Provide each group with several pieces of paper cut to the size of a double business card to allow them to create and revise their ideas for the graphics. Make sure that each group completes a Daily Planning Log to ensure conscious allocation of its resources of time and manpower, as well as to provide you with an assessment of each day's work. Students should use about half of their available time to create graphs for their presentation.

Answers

Answers may vary.
A sample card is shown.

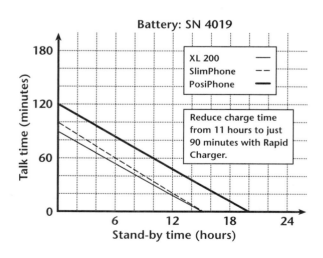

Solving Problems Involving Quadratic Equations

Modern Motors Project

Project Description

The goal of this project is to have students solve problems involving quadratic equations, organize their results, and communicate them in an easily interpreted format within a predetermined time period.

Life/Work Skills

- Identifying an appropriate problem-solving strategy
- Acquiring necessary information
- Communicating effectively by using charts and graphs
- Working as part of a team

Introducing the Project

You may want to review quadratic equations and how to convert measurements by using a constant. Prepare a few of your own charts or graphs as samples of high quality work. Ask groups to examine your samples and to answer the question, What makes an effective chart or graph?

Time Line 3–5 days

Select a definite number of days that students can spend on their projects to ensure that they manage their time. Students should spend their first day seeking an understanding of the problem and determining the subskills required to create a finished product. Make sure that each group completes a Daily Planning Log to ensure conscious allocation of its resources of time and manpower, as well as to provide you with an assessment of each day's work. Students should use about half of their available time to create charts and graphs for their presentation.

Answers

Answers may vary. A sample chart is shown.

Stopping Information for Modern Motors Class-1

Speed (mph)	Reaction distance (ft)	Braking distance (ft)	Stopping distance (ft)	Standard stopping distance (ft)
10	10	4.5	14.5	15
20	20	18.2	38.2	40
30	30	40.9	70.9	75
40	40	72.7	112.7	120
50	50	113.6	163.6	175

Using Surface Area to Solve Problems

Reclamation Project

Project Description

The goal of this project is to encourage students to use the given information to generate a conclusion by discovering and extending patterns and using critical-thinking skills.

Life/Work Skills

- Identifying an appropriate problem-solving strategy
- Distinguishing between meaningful and extraneous information
- Acquiring necessary information
- Working as part of a team

Introducing the Project

You may want to review how to find the surface area of a sphere and how to draw an accurate model from given information. Prepare a few of your own models as samples of high quality work. Ask groups to examine your samples and to answer the question, What makes an effective model?

Time Line *3–5 days*

Select a definite number of days that students can spend on their projects to ensure that they manage their time. Students should spend their first day seeking an understanding of the problem. They should draw a preliminary model of the site and experiment with different ideas for the positions of the trees. Make sure that each group completes a Daily Planning Log to ensure conscious allocation of its resources of time and manpower, as well as to provide you with an assessment of each day's work. Students should use about half of their available time to create their model and displays for their presentation.

Answers

Answers may vary.

Straiter's Stairs Project

Daily Planning Log   Day ___1___

As a group, state what you plan to accomplish and what information and resources you will need.

We need to calculate stair dimensions for four different heights by using the "Pythagorean" Right-Triangle Theorem. We need grid paper, a ruler, and a calculator for the diagrams.

Group members	Roles	Responsibilities and due dates
Charlie	presenter	present to class, help with math
Maria	calculator	perform calculations
Lee	artist	draw diagrams, help with math
Darrell	labeler	label diagrams, help with math • diagrams finished by Thursday • presentation on Friday

What did your group accomplish today? *read and understood problem, established roles, and distributed workload*

What tasks remain to be done? *mathematical calculations, diagrams, and presentation*

Where will you begin tomorrow? *start calculations for 3-foot staircase*

Assessment of today's work	Comments
0 1 ② 3 4 The group stayed on task.	*Once we got started, we did well.*
0 1 2 ③ 4 Each member contributed.	
0 1 2 3 ④ Each member had clearly stated roles.	*See the chart above.*
0 1 2 3 ④ The group respected each member.	*We all got along well.*
0 1 2 ③ 4 Everyone learned something.	

PRESENTATION EVALUATION RUBRIC FOR LONG-TERM PROJECTS

GROUP _____ PROJECT _____ SCORE _____

	UNSATISFACTORY Score 0	ADEQUATE Score 1	EFFECTIVE Score 2	EXEMPLARY Score 3
REASONING Score ___	• little or no reasoning demonstrated	• some logic, but reasoning was not thorough or clearly expressed	• logical reasoning expressed most of the time	• good reasoning and logic are clearly evident
MATHEMATICAL UNDERSTANDING Score ___	• no or inappropriate use of mathematical language and symbolism	• some appropriate use of mathematical language and symbolism	• mostly appropriate use of mathematical language and symbolism	• precise and appropriate use of mathematical language and symbolism
	• inappropriate or incomplete graphs, tables, diagrams, or models	• mostly accurate graphs, tables, diagrams, or models, but they are mislabeled or unclear	• accurate graphs, tables, diagrams, or models with appropriate labels	• sophisticated graphs, tables, diagrams, or models with clear and accurate labels
PRESENTATION Score ___	• presentation lacks organization and coherence	• routine presentation with only minor lapses in organization or clarity	• interesting presentation with good organization and coherence	• clear, well-organized, and engaging presentation
	• none of the requirements of the memorandum are completely communicated to the intended audience	• some of the requirements of the memorandum are completely communicated to the intended audience	• all requirements of memorandum are addressed, and most are effectively communicated to intended audience	• all requirements of memorandum are completely and concisely communicated to intended audience
FEASIBILITY Score ___	• inappropriate or unworkable approach	• approach is workable some of the time	• practical approach	• efficient or sophisticated approach
ORIGINALITY Score ___	• no attempt at originality	• attempt at originality evident	• clear extension of standard approach	• original and resourceful approach

Tech Prep Applications
Construction, Chapter 1

PROFILE: Workers in the construction field construct, alter, and maintain buildings and other structures as well as operate various machines. The construction industry, one of the economy's largest industries, boasts a very large number of self-employed workers. Most construction firms employ fewer than 10 people. Most job opportunities in this field require a high school diploma. However, many desirable jobs, such as construction managers, require a Bachelor of Arts degree in Construction Science with an emphasis on Construction Management.

JOB SEARCH OUTLOOK: The number of jobs in the construction industry should grow about 26 percent from 1996–2006. The fastest growing jobs include bricklayers, electricians, sheet-metal workers, duct installers, drywall installers, painters, and heating, air-conditioning, and refrigeration technicians. Growth is also expected in the construction and repair of streets, bridges, and highways.

A builder is developing a residential housing community called Westridge with eight different model homes.

1. The Sundown model has a tile entryway. The pattern is made up of five different tiles. The first four are shown in the diagram. Draw the next two tiles in the pattern.

2. The prices of the homes in Westridge vary according to the model. Based on the data given, predict the prices of the next three homes. Complete the table.

Model name	Price
Sundown	$109,900
Pinery	$111,700
Westbrook	$113,500
Bayfield	$115,300
Stony Creek	$117,100
Meadows	
Mt. Evans II	
Riverview	

Tech Prep Applications
Construction, Chapter 1, page 2

Write an equation for each situation, and solve it to answer the question.

3. There are 8 Bayfield models in Westridge. Each house will need 7 doors for rooms and 6 doors for closets. How many doors will be needed for all of the Bayfield models?

_____

4. As part of the purchase price of the Stony Creek model, the builder has agreed to wallpaper the family room. It takes 12 rolls of wallpaper to cover the family room walls. If it costs the builder $288 to wallpaper one room, how much does each roll cost?

_____

5. A contractor ordered 66 spools of electrical wire. If the total amount of wire ordered equals 2970 feet, how many feet of wire are on each spool?

_____

6. The builder estimates that it will take 8 days to do the pipe installation in one home. If a plumber works 8 hours per day and earns $1216 for the entire job, what is the plumber's average hourly rate of pay?

_____

The builder hires several framers to build the houses in the Westridge development. Their wages vary depending on how much overtime they work. Complete the wage worksheet below.

	Employee	Hours worked	Rate $8 per hour, plus $12 per hour over 40 hours	Total wages
			WAGE WORKSHEET	
7.	Garcia	45	8(40) + 12(5)	$380
8.	DiMaria	42		
9.	Chin	53		
10.	Walton	48		
11.	Johnson	60		
12.	Brock	59		

Tech Prep Applications
Construction, Chapter 1, page 3

The cost of renting a piece of equipment is given by the formula $c = 20 + 30h$, where c is the total cost and h is the number of hours the equipment is rented.

13. Make a table to show the total cost of renting the equipment for 1, 2, 3, 4, and 5 hours.

14. On the grid provided, make a bar chart of the data in your table.

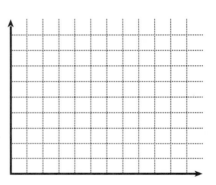

15. Use the bar chart to determine how many hours the builder can rent the equipment if there is $145 in the budget.

_____

16. The builder received a bill from the equipment rental company for $260. How many hours did the builder rent the equipment?

_____

17. How much would it cost the builder to rent the equipment for 24 hours?

_____

18. A different rental company rents the same equipment at a flat rate of $695 per day. How does this compare to the 24-hour rate for the other rental company?

_____

Tech Prep Applications
Construction, Chapter 1, page 4

A construction firm estimates that the cost of building new houses is approximately $38 per square foot. The price charged to the consumer is approximately $45 per square foot. The difference between the costs and the price charged to the consumer is the company's profit. Use the table below for Exercises 19–24.

Size of house in square feet	Costs of construction in dollars	Price to customer in dollars	Company's profit in dollars
1500			
2000			
2500			
3000			
3500			

19. Write an equation to represent the cost of construction, c, for a house that has an area of n square feet. _____

20. Use the equation that you wrote in Exercise 19 to complete the second column in the table.

21. Write an expression to represent the price to the consumer, p, for a house that has an area of n square feet. _____

22. Use the equation that you wrote in Exercise 21 to complete the third column in the table.

23. Complete the last column in the table by finding the profit that the company will make on each house.

24. If the company had to choose between a contract to build a 2000-square-foot house and a contract to build a 3500-square-foot house, which contract would you recommend that they choose? Explain your answer.

_____

_____

_____

_____

Tech Prep Applications
Meteorology, Chapter 2

PROFILE: Meteorology is divided into two major branches, dynamic and synoptic. Dynamic meteorology deals primarily with the motions of the atmosphere and the physical processes involved in air flow. Research in the field involves the extensive use of computer models of general global circulation and of small-scale motion systems such as tornadoes and hurricanes. Synoptic meteorology concentrates on atmospheric phenomena that are directly associated with weather.

JOB SEARCH OUTLOOK: The employment outlook for meteorologists from 1996 to 2005 is about average. Many colleges and technical school offer a two-year program in meteorology. Some meteorologists learn some of their skills on the job. Employment is available through government agencies or private companies such as airlines.

Meteorologists record temperatures, wind velocities, air pressure, and humidity.

1. On Sunday night, the lowest temperature was −7°F. On Wednesday night, the lowest temperature was −12°F. Which night was colder? By how many degrees?

   _____

2. A meteorologist measured the air pressure every hour. Each hour the change in pressure was −8 millibars. What was the total change after 4 hours?

   _____

3. Remote-sensing systems such as balloon-borne radiosondes collect weather data. A balloon at an altitude of 70,000 feet measured the temperature as −57°F. The temperature at ground level was 78°F. How much warmer was it at ground level?

   _____

4. The wind velocity changed from 31 mph to 23 mph. Use a signed number to express the amount of the change.

   _____

5. At noon, the temperature was 15°F. At 6:00 P.M., it was 17 degrees lower. What was the temperature at 6:00 P.M.?

   _____

Tech Prep Applications
Meteorology, Chapter 2, page 2

6. The temperature dropped from 0°C to −36°C in 6 hours. What was the average change per hour?

_____

WINDCHILL

When you lose body heat, you feel cold. Wind can cause a loss of body heat. On a cold, windy day you could feel colder than the actual temperature because of the wind.

Windchill factor is a measure of the cooling power of the air in relation to temperature and wind speed. Humidity is not considered when calculating the windchill factor. Sweating and rapid movements such as downhill skiing increase the cooling power, whereas bright sunshine decreases it.

The table shows windchill factors in degrees Fahrenheit.

Temperature (°F)

Wind speed (mph)	35	30	25	20	15	10	5	0	−5	−10	−15	−20
5	33	27	21	16	12	7	0	−5	−10	−15	−21	−26
10	22	16	10	3	−3	−9	−15	−22	−27	−34	−40	−46
15	16	9	2	−5	−11	−18	−25	−31	−38	−45	−51	−58
20	12	4	−3	−10	−17	−24	−31	−39	−46	−53	−60	−67
25	8	1	−7	−15	−22	−29	−36	−44	−51	−59	−66	−74

Example: The temperature is 20°F. The wind is blowing at 10 mph. How cold does it feel to your body?

Solution: Look at the table to find the column for 20°F. Look down the left column to find the row for 10 mph. At the intersection of the column and the row is the answer, 3°F.

7. Find the windchill factor if the actual temperature is 30°F and the wind

speed is 25 mph. _____

Tech Prep Applications
Meteorology, Chapter 2, page 3

8. What is the windchill factor if the actual temperature is −5°F and the

wind speed is 5 mph? _____

9. If the windchill factor is −31°F and the wind speed is 15 mph, what is

the actual temperature? _____

10. If the windchill factor is −15°F and the actual temperature is 20°F,

what is the wind speed? _____

THERMOMETERS

The thermometer shows the temperature outside of Francie's house on a
cold winter day.

11. How is the thermometer similar to a number line?

_____

12. Using a colored pencil, show an increase of 5 degrees on the
thermometer.

_____

13. Using a different colored pencil, show a decrease of 2 degrees on the
thermometer.

_____

14. By how many degrees must the original temperature increase to reach 0°F.

_____

15. What other temperature is the same distance from 0°F as the one
shown?

_____

16. By how many degrees would the temperature need to increase before
it would be off the scale of the thermometer?

_____

Tech Prep Applications
Meteorology, Chapter 2, page 4

PROJECT

Synoptic meteorologists use data from ground-based radar and remote-sensing systems to make short-range forecasts of local weather. Such forecasts project atmospheric conditions for time periods ranging from a few hours to 12 hours. Long-range predictions employ numerical weather prediction models, consisting of mathematical representations of the laws of motion, heat, mass, and moisture.

The use of high-speed supercomputers makes it possible to predict conditions fairly accurately five to seven days in advance. Research the technology used in the field of meteorology. Write a description of each device and explain how it works. You may wish to draw a sketch of each instrument or device to illustrate your description. Explain how numerical forecasting is used in weather predictions. Use additional paper, if necessary.

_____

_____

_____

_____

_____

_____

_____

_____

_____

_____

_____

_____

_____

_____

_____

_____

_____

Advanced Technological Solutions

TO: Algebra Consulting Teams
FROM: President of ATS
RE: AirTech Aircraft Project

Background

AirTech Aircraft Industries is in the final stage of preparing its new airplane, the AirTech IV, for production. An independent laboratory furnished AirTech with the attached performance information. We have been contracted to analyze the performance information and to present this important information in an easy-to-understand format. The documentation of our results should be arranged in tables and graphs suitable for inclusion in the Pilot Operating Handbook, which is included with each plane when it is delivered to a customer.

Task

AirTech wants us to provide them with performance documentation for flight altitudes of 2000, 3000, 5000, 8000, and 12,000 feet under each of the following conditions:

- the travel range on one tank of fuel with no wind and no reserve fuel
- the travel range on one tank of fuel with a 10, 20, and 30 knot headwind and no reserve fuel
- the travel range on one tank of fuel with a 10, 20, and 30 knot tailwind and no reserve fuel
- the travel range on one tank of fuel with no wind and 45 minutes of fuel withheld as a reserve
- All ranges should be given in nautical miles (1 knot = 1 nautical mile per hour).

Cruise Power Settings for AirTech IV
(Usable fuel is 48 gallons.)

Altitude	Outside temperature		Engine speed		Air speed	
(°Ft)	(°F)	(°C)	(RPM)	(gal/h)	(knots)	(mph)
0	63	17	2400	11.5	150	173
2000	55	13	2400	11.5	153	176
4000	48	9	2400	11.5	156	180
6000	41	5	2400	11.5	158	182
8000	36	2	2400	11.5	161	185
10,000	28	−2	2400	11.5	163	188
12,000	21	−6	2400	10.9	163	188
14,000	14	−10	2400	9.9	160	184

AirTech Aircraft Project

Daily Planning Log   *Day* ____

As a group, state what you plan to accomplish and what information and resources you will need.

_____

_____

_____

Group members	Roles	Responsibilities and due dates

What did your group accomplish today?_____

_____

What tasks remain to be done? _____

_____

Where will you begin tomorrow? _____

_____

Assessment of today's work	Comments
0 1 2 3 4 The group stayed on task.	
0 1 2 3 4 Each member contributed.	
0 1 2 3 4 Each member had clearly stated roles.	
0 1 2 3 4 The group respected each member.	
0 1 2 3 4 Everyone learned something.	

Tech Prep Applications
Landscape Architecture, Chapter 3

PROFILE: Landscape architecture is one of the decorative arts, similar to architecture, city planning, painting, sculpture, and horticulture. Landscape architects work to make outdoor areas beautiful and practical for use. Landscape architects begin with the natural terrain and enhance, recreate, or alter pre-existing landforms. Landscapers use trees, flowers, water, and rocks as well as man-made structures such as decks, terraces, plazas, gazebos, and fountains. The sizes of projects may vary from household gardens to parks, campuses, and urban areas. Design styles include classical/symmetrical, natural/romantic, private/public, and formal/informal.

JOB SEARCH OUTLOOK: The employment outlook for landscape architects from 1996–2005 is average. A bachelor's degree is required and advancement opportunities appear to be better than average. Although the turnover rate in this field is higher than average, the median salary is $41,900 per year.

The Treeline Company is creating the landscape for a new office building.

1. Treeline has designated two-fifths of its landscaping budget for planting new trees. What percent of the budget is reserved for new trees? _____

2. If the amount budgeted for new trees is $4500, how much is the total budget for landscaping? _____

3. If Treeline plans to buy 80 trees with the budgeted amount, what is the average price of each tree? _____

4. Two-thirds of the project is projected to be completed in 4 weeks. At this rate, how long will the project take from start to finish? _____

5. One area of the landscape is a grassy region in the shape of a rectangle. To determine how much sod is needed to cover the area, Treeline must know the area of the region. If the region measures 36.8 feet by 22.4 feet, what is the area of the grassy region? _____

6. Treeline wants to apply a fertilizer to the grassy area. Treeline makes its own fertilizers by combining nitrogen, phosphate, and potassium in specified ratios. For this lawn, they need a ratio of 5:2:4. The worker at the fertilizer center has prepared three batches of fertilizer marked with the weight, in pounds, of each ingredient. The batches are labeled 15:9:5, 15:6:12, and 12:5:8. Which batch has the desired ratio of 5:2:4? Explain.

_____

_____

Tech Prep Applications
Landscape Architecture, Chapter 3, page 2

Treeline is also marketing a new pesticide that contains rotenone for use on lawns. Rotenone is a natural pesticide that is found in the roots of some tropical plants. Treeline sells the concentrated pesticide as a liquid that is mixed with water in the ratio of $1\frac{1}{2}$ ounces of rotenone per gallon of water.

7. If 1 gallon is equivalent to 128 ounces, express the ratio of pesticide to water in lowest terms.

_____

8. Suppose that you want to make the pesticide mixture with only 1 quart of water. How much pesticide would you need?

_____

9. If the automatic sprayer is preset ot mix 5 gallons of water, how much pesticide should be added to the sprayer?

_____

10. If each gallon of the pesticide mixture covers approximately 800 square feet, how much area would be covered by 5 gallons of the mixture?

_____

11. How many times will the automatic sprayer need to be filled to cover an area of 12,000 square feet?

_____

Landscape architects often use custom-made or commercial software packages to prepare their designs. Research the benefits of using a computer to design an area to be landscaped. Call local landscape design firms and ask how they use available technology on their contract projects. Preview a software program or call the manufacturer and find out the program's features. Write a summary of your findings.

_____

_____

_____

_____

_____

_____

_____

_____

_____

_____

Tech Prep Applications
Business Services, Chapter 3, page 3

PROFILE: One of the fastest growing business services—and one of the fastest-growing industries in the economy—will continue to be computer and data processing services. This industry's rapid growth stems from advances in technology, worldwide trends toward office and factory automation, and increases in demand from business firms and individuals. Personnel supply services, primarily temporary help agencies, make up the fastest-growing business service industry.

JOB SEARCH OUTLOOK: Business services industries will generate many jobs. Employment is projected to grow from 5.2 million in 1990 to 7.6 million in 2005.

Johnny is one of 32 employees that work for Quality Personnel. Quality has been asked to provide 8 office assistants to an accounting firm and 2 receptionists for a dental office. The employees for both assignments are randomly chosen by a computer.

1. What is the probability that Johnny will be chosen to work for the accounting firm?

   _____

2. What is the probability that Johnny will be chosen to work for the dental office?

   _____

3. What is the probability that Johnny will be assigned to either one of the businesses?

   _____

4. After the first person is chosen, does the second choice represent a dependent or an independent event? Explain.

   _____

   _____

5. Just before the selections were made, the accounting firm called and requested that 4 more office assistants be sent over. How will this change affect the probabilities in Exercises 1–3? Explain.

   _____

   _____

6. Johnny's manager told him that he could expect to work about 60% of the workdays during the winter. How many days should Johnny expect to work in the next 6 weeks? Assume that Johnny works only Monday through Friday.

   _____

Tech Prep Applications
Business Services, Chapter 3, page 4

The committee conducted an opinion poll to find out if the employees of the company favor a 4-day work week. Five divisions were polled. The results of the opinon poll are shown below.

Find the number of people who responded favorably in each office.

7. California office: 42% of 300 people surveyed _____

8. New York office: 28% of 350 people surveyed _____

9. Seattle office: 55% of 460 people surveyed _____

10. Austin office: 34% of 250 people surveyed _____

11. New Orleans office: 63% of 500 people surveyed _____

Collecting data from opinion polls and surveys is more efficient when technology is used. Databases and random number generators make it easy to collect samples and make predictions.

Design a survey to be taken by using technology. Explain how you decided to use software and technology to make the process more efficient. Write a detailed explanation of your plan. Explain why you think your method saves time and is more accurate than doing it manually. Write a summary of your findings.

_____

_____

_____

_____

_____

_____

_____

_____

_____

_____

_____

_____

Tech Prep Applications
Photography, Chapter 4

PROFILE: Photography is the method of recording the permanent image of an object by the action of light, or related radiation, on a sensitive material. Publicity, fashion, and advertising photography have become highly sophisticated, as have industrial and architectural photography. Photography is also used for scientific and technical research. It has made a direct contribution to work in virtually every discipline.

JOB SEARCH OUTLOOK: The employment outlook for photographers from 1996 to 2005 is average. A bachelor's degree is required, and advancement opportunities appear to be better than average. About half of all photographers are self-employed.

Technical improvements in the camera have transformed it from a bulky, cumbersome apparatus to a compact, sophisticated device that is often small enough to fit into a pocket.

A photographer uses a variety of lenses depending on the object being photographed. A wide-angle lens is often used to take photos of landscapes and panoramic views. A telephoto lens is used to take magnified photos.

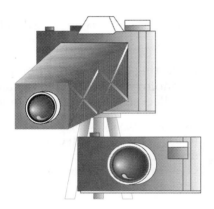

Measure the field of view (angle) of each lens.

1.

_____

2.

_____

3.

_____

Tech Prep Applications
Photography, Chapter 4, page 2

4. A photographer takes a photo of a building. One centimeter on the picture represents seven meters on the building. What is the scale of the photo?

_____

5. A photographer takes a photo of a maple leaf. Three inches on the photo represents one inch on the leaf. What is the scale of the photo?

_____

6. Suppose that a photographer wants to frame a photo. Each of the four frame pieces is 12 inches long. She wants to measure the diagonals to make sure that the frame is square. How long should the diagonal measurement be (at its longest point) for this frame?

_____

PROJECT

Computer technology has revolutionized the way that photographs are used in magazines, newspapers, and catalogs. Software programs are available that enable the user to scan color photographs into a graphics program and alter the photo on the screen by using a computer.

Research available programs such as the one described. Write a detailed report, summarizing how it works and how it is used in the business world . Include the special features of the software and explain how it works in conjunction with other programs such as word-processing software. In addition, be sure to include any new technologies in development that would improve on what is already available.

_____

_____

_____

_____

_____

_____

_____

_____

_____

Tech Prep Applications
Civil Engineering, Chapter 4, page 3

PROFILE: The profession of civil engineering can be divided into three broad categories: consulting, contracting, and municipal engineering. The consulting engineer is the technical advisor to the client, who may be an individual, a commercial firm, or a government department. The contracting engineer is responsible for surveys, designs, specifications, and schedules for the project, as well as supervision of the construction. The municipal engineer serves local or state governments directly through the planning and construction of water supply and sewage disposal systems, roads, bridges, and other public facilities.

JOB SEARCH OUTLOOK: The employment growth for civil engineers from 1996 to 2005 is average. A bachelor's degree is required, and the median starting salary is $29,376.

In a section of a new neighborhood called Park Meadows, the city blocks are arranged like a grid. The roads running east and west are called avenues and the roads running north and south are called streets.

1. What is true of all of the avenues in Park Meadows?

_____

2. What is the measure of the angle formed when a street intersects an avenue? Explain.

_____

3. What shape are the city blocks? Explain how you can identify their shape.

_____

4. In another neighborhood, the streets intersect at different angles. Use the space provided to make a diagram of two streets that intersect at an angle of 75°. Each street is 30 feet wide, but at the intersection, the distance from corner to corner is 31 feet.

5. If the angle at one corner is 75°, what is the

measure of the angle at the other corner? _____

6. Draw a right triangle, using the distance between two corners as the hypotenuse and the distance straight across one of the streets as one of the legs. Label and measure the angles within the right triangle.

Tech Prep Applications
Civil Engineering, Chapter 4, page 4

7. Find the measure, to the nearest foot, of the unknown side of the right triangle that you drew in Exercise 6.

_____

8. What kind of polygon is formed by the intersection of the two streets?

_____

9. An engineer is preparing to repave this intersection. Find the area of the region formed by the intersection.

_____

PROJECT

Civil engineers are capable of interpreting a design and devising methods for carrying out the work. They organize and control both labor and machines. The duties of the contracting civil engineer are to make detailed surveys of site conditions, obtain information about the supply of materials, plan in detail how the work shall be carried out, and determine what quantity of both machines and labor will be required.

Research the procedure that a civil engineer must follow in planning and developing new streets in a new division of a city. Using manual or computer-aided drafting equipment, create a design of the area. Summarize your findings in a short essay.

_____

_____

_____

_____

_____

_____

_____

_____

_____

_____

Advanced Technological Solutions

TO: Algebra Consulting Teams
FROM: President of ATS
RE: Straiter's Stairs Project

Background

The employees of Straiter's Carpentry, Inc. have experienced difficulty building stair stringers, the boards which support the steps. Stair stringers are cut from 2-inch-by 12-inch-boards, and employees have ruined many boards because of making one cut in the wrong place. A sample stringer is shown at right.

We have been contracted to design templates for the stringers according to certain specifications they have provided. Each stringer should maintain as much of the 12-inch width as possible to ensure maximum support.

The templates should be scale drawings that include the dimensions of the board and the length of each cut. They should be easy to read and suitable for inclusion in the handbook that is presented to each employee during training.

Task

Straiter's Carpentry, Inc. wants us to design four templates for stair stringers according to the following specifications:

- a vertical step up (rise) of no more than 9 inches
- horizontal and vertical components that add up to 17 inches
- total stair heights of 3 feet, 4 feet, 6 feet, and 8 feet

TECH PREP LONG-TERM PROJECT

Straiter's Stairs Project

Daily Planning Log   Day ____

As a group, state what you plan to accomplish and what information
and resources you will need.

_____

_____

_____

Group members	Roles	Responsibilities and due dates

What did your group accomplish today? _____

_____

What tasks remain to be done? _____

_____

Where will you begin tomorrow? _____

_____

Assessment of today's work	Comments
0 1 2 3 4 The group stayed on task.	
0 1 2 3 4 Each member contributed.	
0 1 2 3 4 Each member had clearly stated roles.	
0 1 2 3 4 The group respected each member.	
0 1 2 3 4 Everyone learned something.	

Tech Prep Applications
Business Management, Chapter 5

PROFILES: The fundamental objective of private for-profit companies is to make a profit for their shareholders or owners, or to increase shareholder value. General managers and top executives work to ensure that their organizations meet these objectives. The educational background of managers varies as widely as the nature of their responsibilities. Many general managers and top executives have a bachelor's degree or higher in liberal arts or business administration.

JOB SEARCH OUTLOOK: Employment of general managers and top executives is expected to grow about as fast as average for all occupations through the year 2005 as new companies start up and established companies seek managers who can help them maintain a competitive edge in domestic and world markets. In addition, because this is a large career field, many openings will occur each year as executives transfer to other positions, start their own businesses, or retire.

A mail-order company is planning to subscribe to an address service. The charge for using the service is a monthly fee of $100 plus $25 per hour. The company does not want to spend more than $400 per month for the address service.

1. Write an inequality that describes the relationship between the total cost of the service each month and the maximum amount that the company wants to spend.   _____

2. Write an equation that describes the relationship between the total cost and the monthly and hourly fees.   _____

3. Graph the equation and the inequality on the grid provided.

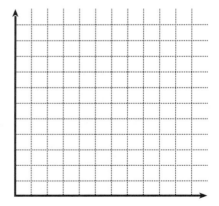

4. Find the maximum number of hours per month that the company should use the service while not spending more than $400.   _____

Tech Prep Applications
Business Management, Chapter 5, page 2

A manager for a company that produces widgets is researching new manufacturing machines. The manager needs to determine the number of new machines to buy for the company. Netus machines can produce 2000 widgets per month and cost $50,000 per machine. Pechel machines can produce 3000 widgets per month and cost $60,000 per machine. The company wants to produce at least 40,000 widgets each month to meet the expected demand. Their budget for machinery purchases is $900,000.

5. Write an inequality for the total cost of buying n Netus machines and p Pechel machines, using the maximum budget of $900,000. _____

6. Write an inequality that expresses the total number of widgets produced with n Netus machines and p Pechel machines, using the minimum of 40,000 widgets. _____

7. Use the grid provided to graph the pair of inequalities.

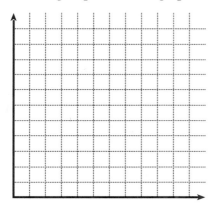

8. Use your graph to determine how many of each brand of machine the manager should buy. Your answer cannot be fractional because whole machines must be purchased. _____

9. If your answer does not meet the expected production demands while minimizing the total cost for the machines, what would your suggestion to the manager be? Explain.

_____

_____

_____

Tech Prep Applications
Agribusiness, Chapter 5, page 3

PROFILE: American farm operators and managers direct the activities of one of the world's largest and most productive agricultural sectors. They produce enough food and fiber to meet the needs of our nation and to export huge quantities to countries around the world.

JOB SEARCH OUTLOOK: Farm operators and managers held about 1,327,000 jobs in 1994. About 90 percent were self-employed farm operators. Most managed crop production activities, while others managed livestock and dairy production. Agriculture employment has been declining for many decades, and this trend will continue. The number of self-employed workers is expected to decline through the year 2005 by 6 percent, from 3.3 million to 3.1 million. However, wage and salary positions will increase by 214,000, with especially strong growth in farm management services.

Joseph manages his family farm. He has a Holstein cow that weighs 850 pounds. The cow eats an average of 24 pounds of feed each day and gains an average of 2.6 pounds per day.

1. Write an equation that expresses the total weight, t, of the cow for each day, d. _____

2. Rewrite the equation for total weight by solving for the number of days, d. _____

3. Solve your equation for Exercise 2 for the number of days that it takes the cow to weigh 900 pounds. _____

4. Write an equation that expresses the number of pounds of feed, f, that the cow eats each day, d. _____

5. Using the equation for f and the number of days calculated in Exercise 13, determine how much feed the cow eats to weigh 900 pounds. _____

Susan also owns a family farm. She is preparing to plant corn and can buy seed corn in $25.00 bags that contain 6000 viable kernels. She plans to plant the seed corn at a rate of 15,000 kernels per acre.

6. How many bags of seed corn are needed for each acre? _____

7. Write an equation for the cost of seed corn, c, per acre, a. _____

8. What is Susan's cost for seed if she plans to plant 55 acres of corn? _____

9. How many bags of corn should Susan order? _____

Tech Prep Applications
Agribusiness, Chapter 5, page 4

Rachel has 48 acres of farmland available. She wants to grow two different varieties of pecan trees to sell as seedlings. Both trees can be planted at 40 trees per acre. A local distributor buys variety A trees for $6 per seedling and variety B trees for $4.50 per seedling. Rachel's distributor will purchase no more than 1000 variety A seedlings and no more than 1200 variety B seedlings.

10. How much can an acre of variety A seedlings cost?   _____

11. How much can an acre of variety B seedlings cost?   _____

12. Write an inequality for the number of acres of variety A seedlings, a, that the distributor is willing to purchase.   _____

13. Write an inequality for the number of acres of variety B seedlings, b, that the distributor is willing to purchase.   _____

14. Write an inequality for the total number of acres that can be planted, using a for the number of acres of variety A seedlings and b for the number of acres of variety B seedlings.   _____

15. If Rachel plants the maximum number of variety A seedlings, how much would she earn from her distributor?   _____

16. If Rachel plants the maximum number of variety B seedlings, how much would she earn from her distributor?   _____

17. Why can't Rachel sell the maximum number of variety A and B seedlings?   _____

18. Using your own paper, construct a graph of acres of variety A seedlings versus acres of variety B seedlings. Graph all of the inequalities on the same coordinate axis and shade the area containing the points that satisfy the conditions given by your inequalities.   _____

19. Using the graph, find the maximum income Rachel can earn from the sale of her seedlings.   _____

Tech Prep Applications
Banking, Chapter 6

PROFILE: Banks are institutions that deal in money and its substitutes and provide other financial services. Banks accept deposits and make loans, deriving a profit from the difference in the interest rates paid and charged, respectively. The banking industry employs tellers, loan officers, managers, loan counselors, computer systems analysts, credit managers, and customer service representatives. The two major classes of banks are commercial and central banks. Commercial banks accept savings and checking deposits, make loans and other investments, and offer financial services that assist in the exchange of funds among individuals and institutions. Central banks act as bankers for governments and as lenders of last resort to commercial banks in cases of financial crisis. Central banks frequently make healthy profits for governments through fees and security transactions.

JOB SEARCH OUTLOOK: The employment rate in the banking industry is expected to grow more slowly than the average for all industries, increasing by only 4 percent over the next decade. Jobs for loan officers and counselors should grow faster than average as applications for commercial, consumer, and mortgage loans increase. Growth in the variety and complexity of loans and the importance of loan officers to the success of banks should ensure rapid employment growth.

Bankers calculate simple interest by using the formula $I = Prt$, where I represents the amount of interest, P represents the principal, r represents the rate of interest given as a percent, and t represents the length of time given in years.

Find the amount of simple interest for each investment. Round your answer to the nearest cent when necessary.

1. $700 at 6% per year for 1 year _____

2. $2500 at 5% per year for 3 years _____

3. $6000 at 12% per year for 4 years _____

4. $600 at $14\frac{1}{2}$% per year for 3 months _____

5. $2000 at 5.4% per year for 6 months _____

Find the rate of interest.

6. $P = \$1200$, $t = 2$ years, $I = \$192$ _____

7. $P = \$600$, $t = 1$ year, $I = \$90$ _____

8. $P = \$6750$, $t = 26$ weeks, $I = \$371.25$ _____

Tech Prep Applications
Banking, Chapter 6, page 2

Find the length of time required for each investment to meet the target interest goal.

9. $2500 at a rate of 10.5% to make $262.50 in interest _____

10. $2000 at a rate of 15% to make $600 in interest _____

11. Ben repaid $2700 on a $2000 loan. The length of the loan was $2\frac{1}{2}$ years. How much interest did he pay? What was the rate of interest?

_____

12. Vanessa borrowed $16,000 to start a small business. She took out two different loans. One loan was for $10,000 at 14%, and the other loan was for $6000 at $15\frac{1}{4}$%. Both loans were for a period of 5 years. How much money must she repay in all?

_____

Banks pay compound interest on savings accounts. Compound interest is paid on both the principal and the interest already earned. Interest can be compounded annually (once a year), semiannually (twice a year), quarterly (four times a year), monthly, or daily.

Suppose that interest on a savings account is compounded semiannually. Follow these steps to find the new balance after 1 year.
• Find the interest after the first payment period.

   Use $I = Prt$. Since interest is compounded semiannually, $t = \frac{1}{2}$ year

• Add the interest to the principal. Find the new principal.
• Find the interest after the next payment period. $t = \frac{1}{2}$ year
• Add the interest to the principal. Find the total principal after one year.

Find the total balance at the end of 1 year. Round your answer to the nearest cent.

13. $4500 at 7% annual interest compounded semiannually _____

14. $900 at 6% annual interest compounded semiannually _____

15. $5000 at 8% annual interest compounded quarterly _____

Tech Prep Applications
Banking, Chapter 6, page 3

Cornerstone Bank has a branch at the airport. Much of the business at the airport branch is concerned with exchanging U.S. dollars for foreign currencies. The exchange is calculated according to the formula $f = rd$, where f is the amount of foreign currency, d represents the amount in dollars, and r represents the exchange rate.

Some common exchange rates are listed in the table.

Exchange Rates (units per dollar)

Country	Unit	Rate (r)
Great Britain	pound	0.59
France	franc	5.05
Germany	mark	1.49
Japan	yen	111.30
Mexico	peso	7.85

Use the formula $f = rd$ to calculate the correct exchange for each requested transaction. Round your answer to the nearest hundredth of a unit.

16. $200 to pesos _____

17. $500 to pounds _____

18. $350 to yen _____

19. $4000 to francs _____

20. $1200 to marks _____

This week there is a large international convention in town, and the airport branch has been very busy exchanging foreign currency for U.S. dollars. In order to exchange foreign currency for dollars, a new formula must be used.

21. Use algebra to solve the equation $f = rd$ for d. _____

Use your formula to calculate the value in U.S. dollars of each of the following amounts. Round your answers to the nearest cent.

22. 2100 francs _____

23. 6800 pesos _____

24. 195 pounds _____

25. 1250 marks _____

26. 8600 yen _____

Tech Prep Applications
Banking, Chapter 6, page 4

PROJECT

The banking industry utilizes the latest technological advancements and automated services to better assist their clients. Banking by computer or using the Automated Teller Machines (ATMs) has been increasingly popular with the public. Video tellers and interactive phone systems are also being introduced into this industry. Research how technology is being used in the banking industry.

The computer is also a very important part of the day-to-day activities of a bank. Contact several local banks in your area and ask about the role that computers play in their day-to-day operations. Write an essay of your findings.

_____

_____

_____

_____

_____

_____

_____

_____

_____

_____

_____

_____

_____

_____

_____

_____

_____

_____

_____

_____

Growth Center Project

Memorandum

TO: Algebra Consulting Teams
FROM: President of ATS
RE: Growth Center Project

Advanced
Technological
Solutions

Background

The Growth Center lawn service is in need of a chart to show the prices that their employees should charge for bulk grass seed. The seed is sold in bags that weigh 5 pounds, 12 pounds, and 25 pounds and in bulk lots over 25 pounds. Each mixture will be a combination of two types of seed: an all-purpose seed will be mixed with a shade seed or a sun seed.

We have been contracted to provide data on mixture amounts and prices for each mixture. They have provided us with the percent of each type of seed to be combined to create each mixture and the prices for each type of seed. The chart that we create should be suitable for inclusion in their employee handbook, which is presented to each employee during training.

Task

The Growth Center lawn service wants us to provide them with data on mixture amounts and prices for each of the following mixtures:

- Economy Sun Mixture   70% all-purpose seed, 30% sun seed
- Economy Shade Mixture   70% all-purpose seed, 30% shade seed

- Standard Sun Mixture   60% sun seed, 40% all-purpose seed
- Standard Shade Mixture   60% shade seed, 40% all-purpose seed

- Deluxe Sun Mixture   80% sun seed, 20% all-purpose seed
- Deluxe Shade Mixture   80% shade seed, 20% all-purpose seed

The Growth Center pays $24 for each 50-pound bag of all-purpose seed, $44 for each 50-pound bag of sun seed, and $32 for each 50-pound bag of shade seed.

The mark-up on the 5-pound bags of mixed seed will be 30% over the wholesale cost of the mixture. The Growth Center will charge a proportionate price for the larger bags but will offer discounts of 5% on a 12-pound bag, 7% on a 25-pound bag, and 9% on bulk orders of more than 25 pounds.

TECH PREP LONG-TERM PROJECT

Growth Center Project

Daily Planning Log   Day _____

As a group, state what you plan to accomplish and what information
and resources you will need.

_____

_____

_____

Group members	Roles	Responsibilities and due dates

What did your group accomplish today?_____

_____

What tasks remain to be done? _____

_____

Where will you begin tomorrow? _____

_____

Assessment of today's work	Comments
0 1 2 3 4 The group stayed on task.	
0 1 2 3 4 Each member contributed.	
0 1 2 3 4 Each member had clearly stated roles.	
0 1 2 3 4 The group respected each member.	
0 1 2 3 4 Everyone learned something.	

Tech Prep Applications
Traffic Engineering, Chapter 7

PROFILE: Traffic engineers usually receive their educational training in civil engineering. Their main job responsibility is to study the flow of traffic and recommend where to place traffic signs or signals to avoid dangerous road conditions. Traffic engineers attempt to promote the orderly movement of traffic and reduce the frequency of accidents.

JOB SEARCH OUTLOOK: The employment growth outlook for traffic engineers from 1996 to 2005 is average. A bachelor's degree is required, and the median starting salary is $29,376.

Cycle length, which is the time required for a complete sequence of signal indications, normally varies between 30 and 150 seconds. A short cycle length reduces the overall capacity at the intersection, while a long cycle may provoke some impatience among drivers.

The average number of vehicles traveling University Boulevard is 90 vehicles every 12 minutes. The light cycle is 80 seconds. Find the number of vehicles per light cycle.

Step 1   Find the number of light cycles in 12 minutes.

$$12 \times 60 = 720 \qquad \text{Change the minutes to seconds.}$$
$$720 \div 80 = x \qquad \text{Divide by the number of seconds per cycle.}$$
$$x = 9$$

Step 2   Find the number of vehicles per cycle.

$$90 \div 9 = 10$$

The average number of vehicles per light cycle is 10.

Find the average number of vehicles per cycle. Round your answers to the nearest tenth.

1. 120 vehicles per 5 minutes with a light cycle of 100 seconds _____

2. 160 vehicles per 20 minutes with a light cycle of 90 seconds _____

3. 165 vehicles per 15 minutes with a light cycle of 80 seconds _____

4. 200 vehicles per 15 minutes with a light cycle of 75 seconds _____

5. 75 vehicles per 10 minutes with a light cycle of 75 seconds _____

Tech Prep Applications
Traffic Engineering, Chapter 7, page 2

The Greenshield formula is based on the fact that on average it takes 3.7 seconds for the first car to go through an intersection and then 2.1 seconds for each car thereafter. A traffic engineer uses the following formula:

$G = 2.1v + 3.7$, where G represents the green time in seconds and v represents the average number of vehicles traveling in each lane per cycle.

Find the green time for a signal light that averages the following number of vehicles per lane in a given cycle. Use $G = 2.1v + 3.7$:

6. 20 vehicles _____

7. 34 vehicles _____

8. 15 vehicles _____

9. 42 vehicles _____

Use the formula for green time to find the number of vehicles per lane in a given cycle for each of the green times below.

10. 41.5 seconds _____

11. 60.4 seconds _____

12. 28.9 seconds _____

13. 66.7 seconds _____

The yellow time depends on the speed limits. Yellow times are longer at intersections where cars travel at higher rates of speed. Yellow time can be determined by using the formula $Y = 0.05r + 1$, where Y represents the yellow time in seconds and r represents the speed limit in miles per hour.

Find the yellow time for traffic lights on roads with the following speed limits:

14. 45 miles per hour _____

15. 35 miles per hour _____

16. 20 miles per hour _____

17. 40 miles per hour _____

Use the formula for yellow time to determine the speed limit on a street with each of the yellow times below.

18. 3.75 seconds _____

19. 2.5 seconds _____

20. 2.25 seconds _____

21. 4 seconds _____

Tech Prep Applications
Traffic Engineering, Chapter 7, page 3

22. At the intersection of Broadway and Belleview, the light cycle is 90 seconds. The cross street averages 125 vehicles in each lane every 15 minutes. Find the green time for this cross street. Show your work.

_____

23. At the intersection of Caley and Gallup, the light cycle is 120 seconds. A cross street averages 140 vehicles per lane during a 22-minute period. Find the green time for the cross street. Show your work.

_____

24. An intersection has a 150-second signal light cycle. The cross street averages 270 vehicles in each lane during a 27-minute period. Find the green time for the cross street. Show your work.

_____

The light cycle at an intersection is the sum of the green and yellow times for both streets. The city wants to place a signal at the intersection of Washington Avenue and Jefferson Street. A recent study recommended the number of cars per cycle shown in the table below.

	Washington Avenue	Jefferson Street
Number of cars per cycle	15	12
Speed limit (mph)	40	30

Use the data in the table to calculate the green and yellow times for both streets and the total cycle time. Record your results in the table below.

25.

	Washington Avenue	Jefferson Street
Green time, $G = 2.1v + 3.7$		
Yellow time, $Y = 0.05r + 1$		
Cycle time, $C = G + Y$		

Total cycle time	

Tech Prep Applications
Traffic Engineering, Chapter 7, page 4

PROJECT I

Time a traffic light in your community during a peak traffic time. Find the average number of vehicles in each cycle. Use Greenshield's formula and compare your result with the actual length of green time.

PROJECT II

Digital computers have made possible more complete and more extensive forms of control. Computer control of large networks of roads is generally called area traffic control. Information is obtained by a variety of traffic detectors installed in the street system, that measure the presence of vehicles, traffic flow rates, stop time, speeds, and turning movements. The greater the number of detectors and measurements, the more complex and expensive the system becomes.

Research the ways in which technology plays an important role in traffic control today. What role does the computer play in a traffic situation? What new methods of electronic control are being developed to help traffic situations in the United States today? Write your findings in an essay.

_____

_____

_____

_____

_____

_____

_____

_____

_____

_____

_____

_____

_____

_____

_____

_____

_____

_____

_____

_____

Tech Prep Applications
Electrician, Chapter 8

PROFILE: Electricians are some of the many workers employed in the construction industry. The construction industry, one of the economy's largest industries, boasts a very large number of self-employed workers. Most construction firms employ fewer than 10 people. Most job opportunities in this field require a high school diploma.

JOB SEARCH OUTLOOK: The number of jobs in the construction industry should grow about 26 percent from 1996 to 2006. The fastest growing jobs include bricklayers, electricians, sheet-metal workers, duct installers, drywall installers, painters, and heating, air-conditioning, and refrigeration technicians.

An electrician's labor fee for a service call is closely related to the amount of time he spends on the job. For example, a job that requires two hours of labor will cost the consumer $70. A four-hour job will cost $120.

1. Write the data given in the form of ordered pairs (t, c) with the labor hours as the first component, t, and consumer cost as the second component, c.

2. Use the grid provided to plot the points represented by these ordered pairs.

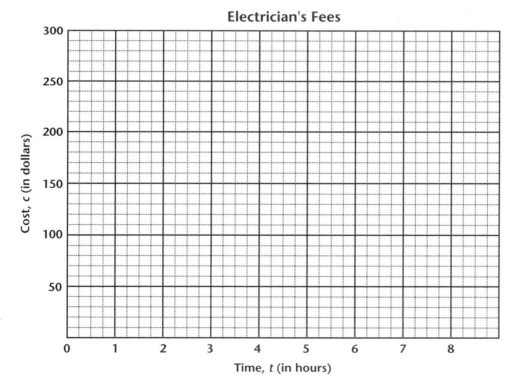

3. Use a straightedge to draw a line through two points on your graph. Extend the line to the edges of the graph.

Tech Prep Applications
Electrician, Chapter 8, page 2

4. The line you have drawn shows the relationship between the electrician's time on the job and the customer's cost. Your line should contain the point (6, 170). What specific information does this ordered pair give concerning the electrician's charges?

_____

_____

Use your graph to approximate the following:

5. the cost of an 8-hour job _____

6. the cost of a 4.5-hour job _____

7. the time spent on a job that costs $195 _____

8. the time spent on a job that costs $110 _____

9. Suppose that the electrician arrives at the home of a customer and the customer informs him that she fixed the problem herself and no longer needs his services. Does the graph indicate that the electrician's fee should be $0? Explain.

_____

_____

10. What is the difference in cost between a 2-hour job and a 3-hour job?

_____

11. What is the difference in cost between a 1-hour job and a 2-hour job?

_____

12. What do you notice about the differences from Exercise 10 and 11?

_____

13. How much does the electrician charge per hour?

_____

Tech Prep Applications
Electrician, Chapter 8, page 3

14. How does the electrician calculate what he should charge for a specific job?

_____

_____

15. Use your answer to calculate the cost for a 3-hour job and a $7\frac{1}{2}$-hour job. Compare your results with the information shown in your graph.

_____

The electrician has charged a $20 truck fee for driving to the consumer's home plus a $25 per-hour labor fee. Suppose that the electrician raises his per-hour fee to $30 but uses the same truck charge. Determine the cost of each of the following service calls. Write the data in the form of an ordered pair.

16. the cost of a 2-hour job _____

17. the cost of an 8-hour job _____

18. the cost of a 5-hour job _____

The graph shows the original time-cost relationship—a $20 truck charge and a $25 per-hour labor charge.

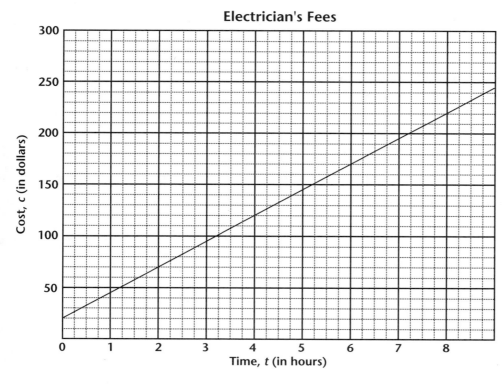

Electrician's Fees

Tech Prep Applications
Electrician, Chapter 8, page 4

19. On the graph, plot the points found in Exercises 16–18 and connect them to show the new time-cost relationship ($20 truck charge and $30 per-hour labor charge).

20. How does the graph show the increase in the per-hour fee?

_____

_____

The graph shows the time-cost relationship for the fees charged by two different electricians, A and B.

21. Which electrician has the higher per-hour fee? Explain.

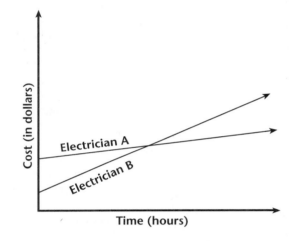

_____

_____

_____

_____

22. Which electrician has the higher truck charge? Explain.

_____

_____

_____

23. Which electrician would you hire if you wanted to spend the least amount of money? Explain.

_____

_____

_____

Mobile Communications Project

Memorandum

Advanced
Technological
Solutions

TO: Algebra Consulting Teams
FROM: President of ATS
RE: Mobile Communications Project

Background

Mobile Communications is in the final stage of completing the package of accessories that comes with their cellular phones. They want to include a 4-inch-by-4.5-inch card that provides information about the amount of stand-by time and the amount of talk time available for each battery type. We have been contracted to create the graphs that will appear on these cards. The card will be folded in half so that it will fit in a pocket or wallet. Each graph should be concise and legible so that the customer can open the card and readily tell how many minutes of talk time are left in the battery after a certain amount of stand-by time has been used.

In addition, they want the recharge time to be shown on the card to give the customer an indication of how much time can be saved if Mobile's Rapid Charger, an optional accessory that can be purchased at a nominal cost, is used. They have provided us with data describing the various batteries and the maximum battery time available for talking and for stand-by.

Task

Mobile Communications wants us to create a series of cards containing graphs based on the maximum times given in the table below. You may assume that the relationship between talk time and stand-by time is linear. Each card should be specific to one battery type and should show the maximum talk and stand-by times for each of the 3 phone types and the information about the Rapid Charger.

Battery type	XL 200 (talk/stand-by)	SlimPhone (talk/stand-by)	PosiPhone (talk/stand-by)	Charge time (overnight/ Rapid Charger)
SN 4019	90 min/ 15 h	100 min/15 h	120 min/20 h	11 h/90 min
SN 4057	150 min/ 24 h	150 min/24 h	180 min/30 h	12 h/90 min
SN 4131	45 min/8 h	55 min/8 h	70 min/11 h	8 h/60 min
SN 4238	130 min/28 h	180 min/28 h	225 min/36 h	11 h/90 min
SN 4258	65 min/10 h	70 min/10 h	90 min/14 h	9 h/60 min

TECH PREP LONG-TERM PROJECT

Mobile Communications Project

Daily Planning Log   *Day* _____

As a group, state what you plan to accomplish and what information and resources you will need.

_____

_____

_____

Group members	Roles	Responsibilities and due dates

What did your group accomplish today? _____

_____

What tasks remain to be done? _____

_____

Where will you begin tomorrow? _____

_____

Assessment of today's work	Comments
0 1 2 3 4 The group stayed on task.	
0 1 2 3 4 Each member contributed.	
0 1 2 3 4 Each member had clearly stated roles.	
0 1 2 3 4 The group respected each member.	
0 1 2 3 4 Everyone learned something.	

Tech Prep Applications
Medical Laboratory Technician, Chapter 9

PROFILE: Medical Laboratory technicians generally have an associate's degree from a community or junior college or a certificate from a hospital, vocational school, or one of the Armed Forces. They perform chemical, biological, hematological, immunologic, microscopic, and bacteriological tests in a medical laboratory. They also prepare specimens for examination and use sophisticated equipment to analyze samples and perform tests. This job generally requires substantial knowledge of chemistry, biology, and mathematics.

JOB SEARCH OUTLOOK: Employment of medical laboratory workers is expected to grow about as fast as average for all occupations through the year 2005. The rapidly growing older population will spur demand for medical services because older people generally have more medical problems. Technological changes will have two effects on employment. New, more powerful diagnostic tests will encourage more testing and provide for new jobs. As tests are simplified, however, each worker will be able to perform more tests in a shorter time period. These factors should balance out so that the growth in this field will remain about average.

Salvador has just been offered a job at Med-Test Laboratories. He will be analyzing blood samples. Med-Test offers a choice of two salary plans to its employees. Plan A pays $240 per week plus $0.60 per analyzed sample. Plan B pays $275 per week plus $0.20 per sample.

1. Write a system of equations to represent the weekly salary plans.

   Plan A _____

   Plan B _____

2. Graph the system of equations on the grid provided.

3. Use your graph to estimate the number of samples required for both salaries to be equal.

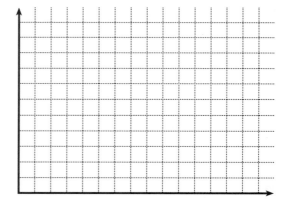

   _____

4. Salvador estimates that he can analyze about 120 samples each week. Which payment plan should he choose? Explain.

   _____

5. Estimate Salvador's weekly salary for the payment plan you selected.

   _____

Tech Prep Applications
Medical Laboratory Technician, Chapter 9, page 2

Another employee at Med-Test Laboratories is preparing a saline, or salt and water, solution. The specifications require 6.5 liters of a 26% solution. The lab stocks pre-mixed solutions of 10% and 50%.

6. Write a system of equations in terms of x and y to calculate the number of liters of each pre-mixed solution that the employee must use to create the specified solution.

_____

_____

7. Solve the system of equations to determine how many liters of each solution the employee needs to use.

number of liters of 10% solution _____

number of liters of 50% solution _____

8. Solve each of the equations that you wrote in Exercise 6 for y, and write them on the lines below.

_____

_____

9. Graph the equations you wrote in Exercise 8 on a graphics calculator. Use the trace feature to find the point of intersection for the two graphs. Does the intersection point verify your answer to Exercise 7? Explain.

_____

_____

10. The doctor has ordered 1 liter of 5% saline solution. Can the employee create this solution? Explain.

_____

_____

11. Another doctor has ordered 3 liters of a 64% saline solution. Can the employee create this solution? Explain.

_____

_____

Tech Prep Applications
Medical Laboratory Technician, Chapter 9, page 3

Med-Test Laboratories has just received a new piece of equipment called a digital pH meter. A pH meter works by relating voltage readings to the pH of a solution. This relationship is nearly linear over the range of pHs from 3 to 12 and is approximated by the following formula:

$P = Va + b$, where P is the pH of a solution,
$\qquad\qquad\qquad$ V is the voltage in millivolts,
$\qquad\qquad\qquad$ and a and b are constants to be determined

The pH meter needs to be calibrated by discovering these constants. The laboratory assistant made two tests of the pH meter that yielded these results.

	pH, P	Voltage, V
First trial	6.5	936 millivolts
Second trial	8.9	1256 millivolts

12. Write a system of equations by substituting the values of P and V from the table into the formula.

_____

_____

13. Solve this system to find the values of the constants a and b.

_____

14. Write a new formula for the relationship between pH and voltage, using the values you found in Exercise 13.

_____

15. Use your formula to find the pH for each voltage.

$V = 1424$ millivolts _____ $\qquad$ $V = 636$ millivolts _____

$V = 1180$ millivolts _____ $\qquad$ $V = 796$ millivolts _____

16. Use your formula to find the voltage for each pH.

$P = 9.2$ _____ $\qquad$ $P = 3.5$ _____

Tech Prep Applications
Medical Laboratory Technician, Chapter 9, page 4

Med-Test Laboratories has decided to expand its business to include a new type of genetic screening. The lab has purchased a new machine to perform the tests. The machine cost $2230, and the materials to process each screening cost $9.50.

17. Write a linear equation that describes the total cost of the screening process in terms of the number of screenings.

_____

18. Graph the equation on the grid provided.

19. Med-Test is planning to charge its customers $65 for each screening. Write a linear equation to describe the total revenue that Med-Test will receive in terms of the number of screenings.

_____

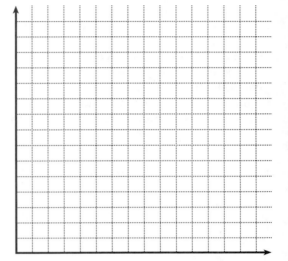

20. Graph this equation on the same grid.

21. Use the graph to approximate the number of screenings that Med-Test will have to perform to break even (to not lose or make money).

_____

22. Use the graph to approximate the amount of profit or loss for each number of screenings.

20 _____   50 _____   100 _____

The head lab technician is scheduling the lab assistants for the next week. There are 60 hours that need to be scheduled, and the company has budgeted $415 per week for lab assistants. The two lab assistants make different wages due their tenure with the company. Rachel earns $7.50 per hour, and Jacob earns $6.50 per hour.

23. Write a system of equations or inequalities that the head lab technician can use to find the maximum number of hours that he can schedule Rachel to work.

_____

_____

24. What is the maximum number of hours that Rachel can work? _____

Tech Prep Applications
Real Estate, Chapter 10

PROFILE: Real estate agents and brokers have a thorough understanding of the real estate market in their communities. They are often independent salespeople who act as a medium for price negotiations between buyers and sellers. Commissions on sales are the main source of income for real estate agents and brokers. Many agents work part time and combine their real estate activities with other careers. Both real estate agents and brokers must be licensed by state agencies and must have received some form of training.

JOB SEARCH OUTLOOK: Employment of real estate agents and brokers is expected to grow as fast as the average for all occupations through the year 2005 due to the growing volume of sales of both residential and commercial properties. Well-trained and hard-working people who enjoy selling have the best chance of success in this competitive field.

In a certain part of town, property values are predicted to increase at a rate of 8% per year for the next 10 years. A client is considering purchasing a piece of property that costs $120,000. The client wants to know how much the property will be worth after 1, 5, and 10 years. She also wants to know how long it will take for the property to double in value.

1. Write a function that will give the value of the property after t years.

   _____

2. What kind of function is the function you wrote in Exercise 1?

   _____

3. Use your function to determine the value of the property at the end of each time period.

   1 year _____

   5 years _____

   10 years _____

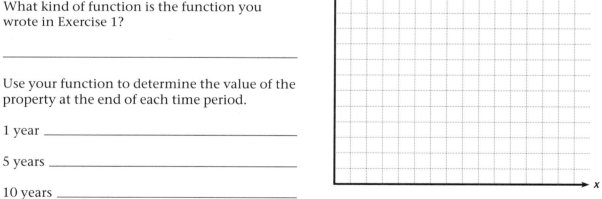

4. Graph the function on the grid provided.

5. Use your graph to estimate when the value of the property will be

   twice the purchase price. _____

Tech Prep Applications
Real Estate, Chapter 10, page 2

6. If the client decided to purchase a piece of land that cost $200,000, would you expect its value to double in the same amount of time? Explain.

_____

_____

7. Use a calculator to verify your answer to Exercise 6.

8. The client is also interested in a $120,000 property in another part of town. Here, property values are expected to remain fixed for the next 5 years until a new freeway is opened. After the freeway is opened, values are expected to rise by 16% per year. How will this investment compare with the original property after 10 years? Which property would you advise the client to buy?

_____

_____

A developer is planning to develop a new subdivision just outside of town. The entire area of the subdivision is 326 acres. The developer wants to set aside 8 acres for a park and recreation center and divide the rest of the land among the residents. He is trying to decide how many lots he should create and how big each lot should be.

9. How big will each lot be if the developer divides the acreage evenly among

100 lots? _____ 200 lots? _____ 300 lots? _____

10. What do you notice about the size of the lots in relation to the number of lots?

_____

11. Write a function for the size of each lot in terms of the number of lots. _____

12. Graph the function on the grid provided.

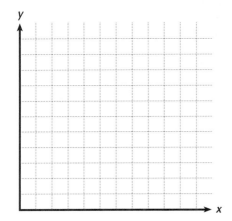

Tech Prep Applications
Real Estate, Chapter 10, page 3

13. If the developer wants the area of each lot to be $1\frac{1}{4}$ acres, approximately how many lots should he create? _____

14. If the developer plans to charge $15,000 for each of the lots described in Exercise 13, how much revenue will he generate if he sells all of the lots? _____

15. If the developer wants to have at least 400 lots in the subdivision, about how many acres will each lot be? _____

16. If the developer plans to charge $9500 for the lots described in Exercise 15, how much revenue will he generate if he sells all of the lots? _____

The ability for a client to obtain a home mortgage depends on the client's income, debt, and credit record. With a good credit record and no long-term debt, the amount of a mortgage that a client can qualify for is shown in the table below.

Annual gross income	$20,000	$30,000	$40,000	$50,000	$60,000
Amount of mortgage (8.0%)	$56,700	$85,100	$113,500	$141,900	$170,300

17. Describe the relationship between annual gross income and the

amount of mortgage that a client can qualify for. _____

_____

18. Use your understanding of the relationship between annual gross income and the amount of mortgage to calculate the amount of mortgage for a $25,000 income. _____

19. What amount of mortgage would you expect a debt-free client with good credit whose gross income is $18,000 to qualify for? _____

Interest rates can be a big factor in the amount of money that a home buyer can qualify for, as shown in the table below for an annual income of $30,000.

Interest rate	6.0%	7.0%	8.0%	9.0%
Amount of mortgage	$104,100	$93,900	$85,100	$77,700

Tech Prep Applications
Real Estate, Chapter 10, page 4

20. Examine the preceding table to determine what type of relationship exists between interest rate and the amount of mortgage. Describe the

relationship. _____

_____

21. Based on your understanding of the relationship between the interest rate and the amount of mortgage, what would be the amount of mortgage for a 10.0% interest rate and an annual income of $30,000?   _____

22. Would you expect the amount of mortgage to increase or decrease if the client had some long-term debt?   _____

23. Most real estate agents charge a commission for their services as a selling agent. This fee is usually a percentage of the sale price of a home. The original homeowner is expected to pay the real estate agent from the proceeds of the sale of their home. Explain what type of relationship exists between the commission percentage and sales price and between the dollar amount charged and the sales price.

_____

_____

Some real estate agents advertise that their commission is a variable percentage rate based on the price of the home being sold. These selling agents charge 5% for homes selling for less than $100,000 and 3% for homes selling for greater than $100,000.

24. What type of function represents this type of variable rate commission?   _____

25. What advantages would a homeowner trying sell his or her home

receive from such an agent? _____

_____

26. Why would some real estate agents want to use a variable percentage rate ?

_____

_____

Modern Motors Project

TECH PREP
LONG-TERM
PROJECT

Advanced
Technological
Solutions

TO: Algebra Consulting Teams
FROM: President of ATS
RE: Modern Motors Project

Background

Modern Motors, Inc. is preparing to unveil its new line of luxury automobiles. Each car comes equipped with anti-lock brakes, which decrease the stopping distance required for speeds up to 50 miles per hour. Modern Motors has hired us to create some charts and graphs that show the comparison between the stopping distance for each of its models and the standard stopping distance.

Stopping distance includes the distance it takes for the driver to react and the distance the car travels once the brakes are applied. The charts and graphs should be easy to read and will be included in a promotional flyer for each model.

Task

Modern Motors is introducing three new automobiles, the Class-1, the Presidential, and the Elite. Our task is to create a chart or graph for each car that shows the reaction distance, braking distance, and stopping distance for the car at speeds of 10, 20, 30, 40, and 50 miles per hour. In addition, each chart or graph should show a comparison to the industry standard.

To calculate the reaction distance, you will need the average driver reaction time (0.68 seconds) and a conversion factor from miles per hour to feet per second (1 mph = 1.47 ft/s). The braking distance of each of these cars and the industry standard are given by the formulas in the table. In the table, v represents velocity in miles per hour and D represents the braking distance required to stop the car in feet.

Model	Class-1	Presidential	Elite	Standard
Braking distance	$D = \dfrac{v^2}{22}$	$D = \dfrac{v^2}{22.5}$	$D = \dfrac{v^2}{24}$	$D = \dfrac{v^2}{20}$

TECH PREP LONG-TERM PROJECT

Modern Motors Project

Daily Planning Log *Day* _____

As a group, state what you plan to accomplish and what information
and resources you will need.

_____

_____

_____

Group members	Roles	Responsibilities and due dates

What did your group accomplish today?_____

_____

What tasks remain to be done? _____

_____

Where will you begin tomorrow? _____

_____

Assessment of today's work		Comments
0 1 2 3 4	The group stayed on task.	
0 1 2 3 4	Each member contributed.	
0 1 2 3 4	Each member had clearly stated roles.	
0 1 2 3 4	The group respected each member.	
0 1 2 3 4	Everyone learned something.	

Tech Prep Applications
Quality Assurance, Chapter 11

PROFILE: Quality assurance technicians monitor quality standards for virtually all manufactured products, including foods, textiles, clothing, glassware, motor vehicles, electronic components, computers, and building materials. They conduct inspections, record the results, compute statistical data, and prepare and present reports on their findings. Quality assurance technicians verify dimensions, color, weight, texture, strength, and other physical characteristics and look for defects or imperfections in the products. Other duties include calibrating precision instruments and contributing ideas for improving the production process. A high school diploma is generally required for jobs in the quality assurance field. Most of these jobs are filled by experienced assemblers, machine operators, or mechanics who already have a thorough knowledge of the products and the production process.

JOB SEARCH OUTLOOK: While the total number of jobs in the quality assurance field is large, employment opportunities in manufacturing are projected to decline over the next 10 years. Quality assurance jobs show an average turnover, giving rise to a large number of openings each year. Most of these jobs are filled by persons with experience in the particular production process involved.

The Audio Ace Company manufactures car stereos. One year ago, the company underwent a management change. The new managers want to compare production totals from this year with those from last year.

1. Use the data provided to make a graph that illustrates the comparison.

Total Production Units by Month

Month	This year	Last year
Jan.	512	541
Feb.	541	630
Mar.	542	492
Apr.	635	665
May	618	645
Jun.	604	598
Jul.	620	576
Aug.	641	525
Sep.	638	518
Oct.	662	511
Nov.	688	496
Dec.	674	510

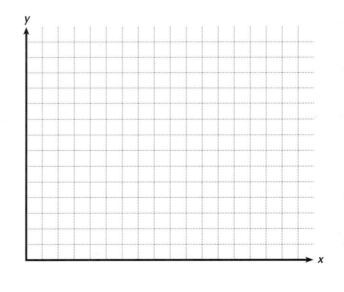

2. For what month was the difference in production between the two years

   the greatest? _____

Tech Prep Applications
Quality Assurance, Chapter 11, page 2

3. What is the range of the data?

_____

4. Find the mean number of units produced each month for this year and for last year.

_____

5. What conclusions can you draw from the data, the graph, and the means that you calculated in Exercise 4?

_____

_____

At the end of each quarter, the company has a shareholders' meeting. The shareholders are very concerned with the number and percent of defective stereos produced each month. Use the data provided to make calculations that will be used in a report for the meeting.

6. The table at right shows the number of defective stereos produced during each week of the quarter. Calculate the mean, median, mode, and range of the data set.

18	20	9
22	24	31
18	17	18
16	18	27

mean _____   median _____

mode _____   range _____

7. Calculate the corresponding percent for the mean, median, and mode if the total number of stereos produced each week is 600.

mean _____   median _____

mode _____

8. Create a stem-and-leaf plot to display the data in the space provided.

9. Calculate the upper and lower quartiles for the data.

upper quartile _____

lower quartile _____

10. Create a box-and-whisker plot to display the data in the space provided.

Tech Prep Applications
Quality Assurance, Chapter 11, page 3

11. The process engineering department has implemented some changes in order to reduce the number of defective components. In a random sample of 150 components, 7 were found to be defective. How does this compare with the data from Exercises 6–10?

_____

12. What recommendations would you make to the process engineering department?

_____

The Audio Ace Company is also interested in identifying the types of defects found in its car stereos. To this end, the quality assurance team has analyzed the defects in a sample of 250 defective stereos. The results are shown below.

- 43 of the components had defective power sources.
- 62 of the components had defective LEDs (light-emitting diodes).
- 73 of the components had mechanical problems.
- 27 of the components had faulty interfaces.
- 15 of the components had defective wiring.
- 30 of the components had exterior defects.

13. Calculate the percent for each type of defect.

power source _____   LEDs _____   mechanical _____

interfaces _____   wiring _____   exterior _____

14. Create a circle graph to display your results.

15. If the total number of defective components last year was 1500, estimate the total number of units with each defect for the year.

power source _____   LEDs _____   mechanical _____

interfaces _____   wiring _____   exterior _____

Tech Prep Applications
Quality Assurance, Chapter 11, page 4

After analyzing the defect report, one of the process engineers suspected that there may be a correlation between the temperature of the machine that installs the power sources and the voltage output of the power source. A random sampling yielded the data shown in the table.

Sample number	Temperature (°F)	Output (volts)		Sample number	Temperature (°F)	Output (volts)
1	78	12.16		11	72	12.71
2	80	11.85		12	74	12.47
3	72	12.64		13	80	11.64
4	72	12.52		14	78	12.10
5	75	12.30		15	71	12.88
6	76	12.25		16	77	12.58
7	79	11.79		17	76	12.34
8	72	12.62		18	73	12.78
9	80	12.01		19	71	12.77
10	73	12.55		20	79	11.83

16. Create a scatter plot to compare the temperature and voltage output for the power supplies.

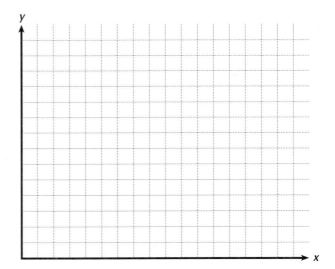

17. Describe the correlation for the data in your scatter plot.

_____

18. If the desired voltage output for a power supply is 12.5 volts or more, what recommendations would you make to the process engineers?

_____

_____

Tech Prep Applications
Aquaculture, Chapter 12

PROFILE: Fishermen generally acquire their occupational skills on the job. Young adults can expedite their entrance into this occupation by enrolling in a two-year vocational-technical program offered by secondary schools, primarily in coastal areas. Fishermen must be in good health and must possess physical strength as well as coordination and mechanical aptitude. Fish culture is the growing of fish in freshwater ponds that allow for feeding, breeding, growing, and harvesting the fish in a well-planned manner. Growing fish in ponds has a 4000-year history, beginning in China. Farmers often add fish ponds to better use their land. For example, fish ponds can be created in a rice paddy or an irrigation system and are very inexpensive to create and maintain.

JOB SEARCH OUTLOOK: Employment of fishermen is expected to decline through the year 2005. Employment growth of fishermen should be restrained by the growing number of large fishing vessels and the use of sophisticated electronic equipment for navigation, communication, and fish location.

Suppose that a farmer wants to create a pond and grow fish to provide food for his family and to earn extra income by selling the surplus fish. Before the farmer can select the types of fish he will raise, he needs to choose whether he will build three small ponds or one large pond. Smaller ponds are easier to drain, refill, and harvest. Larger ponds often take up less space for the same amount of water and cost less to build.

1. Assuming that the farmer wants to build rectangular ponds, complete the table below, which will help him compare ponds of various sizes with depths of 2 feet and 3 feet.

Pond	Size	Volume with 2-ft depth	Volume with 3-ft depth
A	10 ft × 20 ft		
B	10 ft × 25 ft		
C	20 ft × 30 ft		
D	25 ft × 30 ft		
E	25 ft × 35 ft		

2. How much larger is the volume of water in pond C than the volume in pond A for a depth of 2 ft? _____

3. How much larger is the volume of water in pond C than the volume in pond A for a depth of 3 ft? _____

Tech Prep Applications
Aquaculture, Chapter 12, page 2

4. How much larger would you expect the volume of water in pond C to be than the volume in pond A for a depth of 5 ft? Explain your answer.

_____

_____

5. How much larger is the volume of water in pond D than the volume in pond B?

_____

6. How much larger is the volume of water in pond E than the volume in pond B?

_____

7. According to your comparisons, does a larger pond such as pond D occupy less land area than three smaller ponds such as pond A?

_____

_____

The farmer has decided to build a circular pond because circular ponds contain a greater volume of water than rectangular ponds with the same perimeter. To prevent water loss through the bottom of his pond, the farmer wants to use a waterproof plastic liner or clay lining. If clay is selected, a 0.5-meter layer of clay must be put into place.

8. Complete the table to find the surface area of the plastic liner and the cylindrical volume of clay needed for the various sizes of ponds.

Diameter (m)	Surface area (m²)	Volume (m³)
6 m		
8 m		
10 m		
12 m		

9. How much more surface area does a 12-meter diameter pond have than a 6-meter diameter pond?

_____

10. How much more volume does a12-meter diameter pond have than a 6-meter diameter pond of equal depth?

_____

Tech Prep Applications
Aquaculture, Chapter 12, page 3

Very small fish can be difficult to count. If the size of a storage container is known, a small measuring cup can be used to scoop a representative sample of fish and water from the storage container. The fish in the measuring cup can be counted as they are carefully poured back into the storage container. An estimate of the number of fish in the storage container can be calculated using the following equation:

$$\frac{\text{number of fish in cup}}{\text{number of fish in container}} = \frac{\text{volume of cup}}{\text{volume of container}}$$

11. If a 100-mL measuring cup has 20 small fish, how many fish are in a 50-L storage container (50 L = 50,000 mL)? _____

12. If a 200-mL measuring cup has 30 small fish, how many fish are in a 50-L storage container? _____

Pond fertilizers feed the algae in the pond, which are a major food source for the fish. If a farmer wants to add some fertilizer to his pond to improve the pond's productivity, how much should he add? Assume that he selects chicken manure as a fertilizer and that he wants to apply it at the recommended rate of 100 kilograms per 10,000 square meters of pond surface area.

13. If the pond is circular with a diameter of 13 meters, what is the surface area of the pond? _____

14. How much manure should he spread on the surface of the pond? _____

A farmer has purchased a clear plastic cylindrical container that measures 6 feet long and 3.5 feet in diameter. He plans to use it as a nursery pond.

15. How much water does this container hold? Convert your answer from cubic feet to gallons by using the conversion factor 1 cubic foot ≈ 7.5 gallons. _____

16. To help the farmer decide where to store the clear container, he decides to calculate the weight of the container when it is full of water. If 1 gallon of water weighs 6.2 pounds, how much will the full container weigh? _____

17. An interesting feature of the clear plastic container is that its inside surface can receive light, which allows a large surface area for algae growth. Calculate the lateral surface area of the cylinder. _____

18. A circular pond with what diameter would have the same surface area exposed to light as the clear container? _____

Tech Prep Applications
Aquaculture, Chapter 12, page 4

A commonly used fish trap is shaped like a cylinder with a conical entrance, as shown in the figure. Fish enter the trap through the opening at the top of the cone but cannot find the same way out.

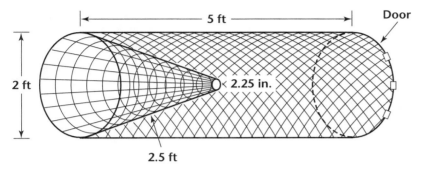

19. Calculate the approximate surface area of the figure in order to determine the minimum amount of wire mesh that is required to build the cage. Show your work.

20. A gill fish net catches fish as they try to swim through the net mesh. If a gill net has square openings of 3 cm by 3 cm to capture mature fish and allow small fish to escape, what is the smallest fish that can be caught in the net? Explain how you arrived at you answer.

_____

_____

_____

Using Surface Area to Solve Problems

Reclamation Project

Memorandum

Advanced
Technological
Solutions

TO: Algebra Consulting Teams
FROM: President of ATS
RE: Reclamation Project

Background
Avoirdupois Mining Company is in the process of planning a reclamation project that will involve planting trees on land previously used for strip mining. Federal regulations require that any land used for strip mining purposes be reclaimed and brought back to a "natural state" by the mining company.

The board of directors at Avoirdupois have decided to try to find a way to turn the reclamation project into a profitable venture. Their research about the Neem Tree has led them to decide to plant, harvest, and market this fast-growing tree for the termite-proof lumber that it yields.

Task
Avoirdupois Mining Company wants us to determine the maximum number of trees that can be planted on the site and to propose a systematic pattern of planting trees to most efficiently utilize the site.

Each tree needs to be at least 24 feet away from any other trees. The site is a hill that is shaped like a hemisphere. The circumference of a great circle of the sphere from which the hemisphere is formed is approximately 452 feet. You are to prepare a presentation that includes a proposed model of the site to present at the next board meeting of Avoirdupois Mining Company.

TECH PREP LONG-TERM PROJECT

Reclamation Project

Daily Planning Log   *Day* _____

As a group, state what you plan to accomplish and what information
and resources you will need.

_____

_____

_____

Group members	Roles	Responsibilities and due dates

What did your group accomplish today?_____

_____

What tasks remain to be done? _____

_____

Where will you begin tomorrow? _____

_____

Assessment of today's work	Comments
0 1 2 3 4 The group stayed on task.	
0 1 2 3 4 Each member contributed.	
0 1 2 3 4 Each member had clearly stated roles.	
0 1 2 3 4 The group respected each member.	
0 1 2 3 4 Everyone learned something.	

ANSWERS

Tech Prep Applications— Chapter 1

Construction

1.
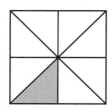 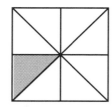

2. Meadows: $118,900
Mt. Evans II: $120,700
Riverview: $122,500

3. $8(7) + 8(6) = 56 + 48 = 104$

4. $24   **5.** 45 feet   **6.** $19/hr

	Employee	Hours worked	Rate $8(40) + 12(h)$	Total wages
7.	Garcia	45	$8(40) + 12(5)$	$380
8.	DiMaria	42	$8(40) + 12(2)$	$344
9.	Chin	53	$8(40) + 12(13)$	$476
10.	Walton	48	$8(40) + 12(8)$	$416
11.	Johnson	60	$8(40) + 12(20)$	$560
12.	Brock	59	$8(40) + 12(19)$	$548

13.

Number of hours, h	1	2	3	4	5
Cost, c		$50	$80	$110	$140	$170

14.
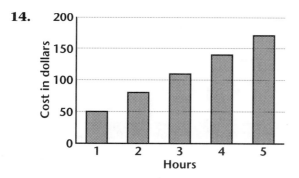

15. 4 hours   **16.** 8 hours   **17.** $740

18. The flat rate of $695 is a better value; it saves the builder $45.

19. $c = 38n$

20.

Size of house in square feet	Costs of construction in dollars
1500	57,000
2000	76,000
2500	95,000
3000	114,000
3500	133,000

21. $p = 45n$

22.

Price to consumer in dollars
67,500
90,000
112,500
135,000
157,500

23.

Company's profit in dollars
10,500
14,000
17,500
21,000
24,500

24. Answers may vary. Sample answer: The company should choose to build the bigger house because the company would make more profit.

Tech Prep Applications— Chapter 2

Meteorology

1. Wednesday was colder by 5 degrees.

2. −32 millibars   **3.** 135 degrees warmer

4. −8 mph   **5.** −2°F   **6.** −6°C per hour

7. 1°F   **8.** −10°F   **9.** 0°F   **10.** 25 mph

11. Answers will vary. Possible answer: They both display positive and negative integers.

12. Check students' drawings.

ANSWERS

13. Check students' drawings.

14. 6 degrees **15.** 6°F **16.** More than 56°F

Check students' projects.

Tech Prep Applications— Chapter 3

Landscape Architecture

1. 40% **2.** $11,250 **3.** $56.25

4. 6 weeks **5.** 824.32 square feet

6. 15:6:12; this ratio can be simplified by factoring out a 3 and leaving the ratio 5:2:4.

7. $\frac{3}{256}$ **8.** $\frac{3}{8}$ of an ounce **9.** 7.5 ounces

10. 4000 square feet **11.** 3 times

Check students' summaries. Be sure they have an understanding of how the technology assists designers in a more efficient way than do paper-and-pencil drawings.

Business Services

1. $\frac{8}{32} = \frac{1}{4}$ **2.** $\frac{2}{32} = \frac{1}{16}$ **3.** $\frac{10}{32} = \frac{5}{16}$

4. The second choice represents a dependent event because the first employee selected cannot be selected again.

5. The probabilities in Exercises 1 and 3 increase to $\frac{5}{16}$ and $\frac{3}{8}$, respectively. The probability in Exercise 2 is unchanged.

6. 18 days

7. 126 **8.** 98 **9.** 253 **10.** 85 **11.** 315

Check students' summaries. They should reflect reasonable thinking and factual information about how technology can be used to collect data.

Tech Prep Applications— Chapter 4

Photography

1. 48° **2.** 16° **3.** 82°

4. 1 cm = 7 m; 1/700 **5.** 3 in. = 1 in.; 3/1

6. about 17 inches

Check students' projects.

Civil Engineering

1. They are all horizontal parallel lines.

2. 90°; since all avenues are horizontal and all streets are vertical, they form perpendicular lines, which intersect at right angles.

3. Rectangles; each block is bounded by two sets of parallel lines that are perpendicular to each other.

4.

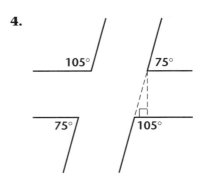

5. 105°

6.

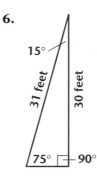

7. 8 feet **8.** parallelogram

ANSWERS

9. 930 square feet

Check students' projects.

Tech Prep Applications— Chapter 5

Business Management

1. $c \leq 400$ **2.** $c = 100 + 25h$

3.

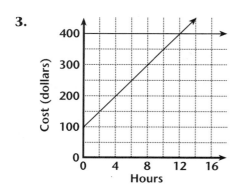

4. 12 hours

5. $50{,}000n + 60{,}000p \leq 900{,}000$

6. $2000n + 3000p \geq 40{,}000$

7.

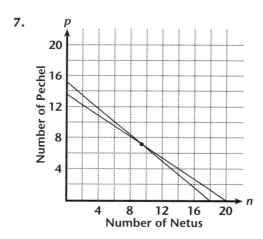

8. Answers may vary because the maximization involves a fraction of a machine. Possible answers are 11 Netus machines and 6 Pechel machines or 10 Netus machines and 7 Pechel machines.

9. Answers may vary. Sample answer: Since a portion of a machine is not possible, 6 Pechel machines should be purchased (instead of 6.67) and an additional Netus machine should be purchased because it is cheaper than an additional Pechel machine. This additional Netus machine will exceed the budgeted amount of $900,000 by $10,000 but will meet the production criteria with the least amount of over expenditure.

Agribusiness

1. $t = 850 + 2.6d$ **2.** $d = \dfrac{t - 850}{2.6}$

3. 19.23 or ≈ 20 days **4.** $f = 24d$

5. ≈ 480 pounds **6.** 2.5 bags

7. $c = 2.5(25)a = 62.5a$ **8.** $3437.50

9. 138 bags **10.** $240 **11.** $180

12. $a \leq \dfrac{1000}{40}$; $a \leq 25$ **13.** $b \leq \dfrac{1200}{40}$; $b \leq 30$

14. $a + b \leq 48$ **15.** $6000 **16.** $5400

17. Rachel does not have enough acreage available to grow the maximum number of seedlings.

18. Check students' graphs.

19. $10,140 for 25 acres of variety A and 23 acres of variety B

Tech Prep Applications— Chapter 6

Banking

1. $42 **2.** $375 **3.** $2880 **4.** $21.75

5. $54 **6.** 8% **7.** 15% **8.** 11%

9. 1 year **10.** 2 years **11.** $700; 14%

12. $27,575 **13.** $4820.51 **14.** $954.81

15. $5412.16 **16.** 1570 pesos

17. 295 pounds   **18.** 38,955 yen

19. 20,200 francs   **20.** 1788 marks

21. $d = \frac{f}{r}$ **22.** $415.84   **23.** $866.24

24. $330.51   **25.** $838.93   **26.** $77.27

Check students' projects.

Tech Prep Applications— Chapter 7

Traffic Engineering

1. 40   **2.** 12   **3.** 14.7   **4.** 16.7   **5.** 9.4

6. 45.7 seconds   **7.** 75.1 seconds

8. 35.2 seconds   **9.** 91.9 seconds   **10.** 18

11. 27   **12.** 12   **13.** 30   **14.** 3.25 seconds

15. 2.75 seconds   **16.** 2 seconds

17. 3 seconds   **18.** 55 mph   **19.** 30 mph

20. 25 mph   **21.** 60 mph

22. 29.95 seconds   **23.** 30.4 seconds

24. 56.2 seconds

25.

	Washington Avenue	Jefferson Street
Green time	35.2 s	28.9 s
Yellow time	3 s	2.5 s
Cycle time	38.2 s	31.4 s

Total cycle time	69.6 s

Check students' projects.

Tech Prep Applications— Chapter 8

Electrician

1. (2, 70) and (4, 120)

2–3. Check students' graphs.

4. The ordered pair (6, 170) indicates that a job which takes 6 hours to complete will cost $170.

5. The cost of an 8-hour job is $220.

6. The cost of a 4.5-hour job is between $130 and $140.

7. 7 hours   **8.** 3.5 hours

9. No; a charge of $20 appears to be made even if no time is spent on the job.

10. a $25 difference   **11.** a $25 difference

12. The differences are the same.

13. The electrician's fee is $25 per hour.

14. The electrician charges $25 per hour plus a $20 truck fee.

15. 3-hour job: $95; 7.5-hour job: $197.50

16. $80; (2, 80)   **17.** $260; (8, 260)

18. $170; (5, 170)

19. Check students' graph.

20. The new line has a greater slope, which indicates an increase in the per-hour charge.

21. Electrician B has the higher hourly rate, as indicated by the steeper slope.

22. Electrician A has the higher truck charge, as indicated by the higher cost of a zero-hour job.

23. Answers will vary. Electrician B would be less expensive for short jobs, but Electrician A would charge less on long jobs. You can see the distinction by looking at the point where the two graphs intersect.

Tech Prep Applications— Chapter 9

Medical Laboratory Technician

1. Plan A: $y = 240 + 0.6x$
Plan B: $y = 275 + 0.2x$

2.

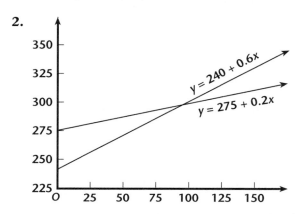

3. about 90 samples

4. plan A because it is the higher salary for any number of samples greater than 90

5. $312

6. $x + y = 6.5$
$0.1x + 0.5y = 0.26(6.5)$

7. 3.9 l; 2.6 l

8. $y = 6.5 - x$
$y = \dfrac{1.69 - 0.1x}{0.5}$

9. The intersection point should verify the answer to Exercise 7.

10. Yes; a 5% solution can be created by adding pure water to one of the other solutions.

11. No; all of the solutions are less than 64%, so they could not be mixed to create a solution that strong.

12. $6.5 = 936a + b$
$8.9 = 1256a + b$

13. $a = 0.0075; b = -0.52$

14. $P = 0.0075V - 0.52$

15. 10.16; 4.25; 8.33; 5.45

16. 1296 millivolts; 536 millivolts

17. $c = 2230 + 9.5s$

18.

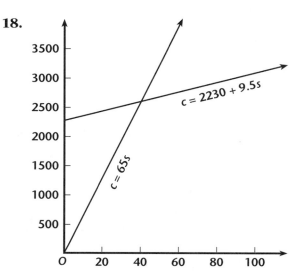

19. $c = 65s$

20. See Exercise 18.

21. Answers may vary but should be close to 41 samples.

22. Answers may vary but should be close to given answers.
loss of $1120; profit of $545; profit of $3320

23. $r + j = 60$
$7.5r + 6.5j \leq 415$

24. 25 hours

Tech Prep Applications— Chapter 10

Real Estate

1. $V = 120,000(1.08)^t$ **2.** exponential

3. $129,600; $176,319.37; $259,071

4.

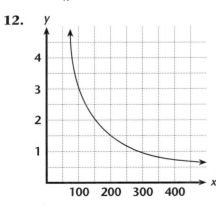

5. shortly after the ninth year

6. Yes; the time that it takes any amount to double depends only on the rate of growth.

7. $200,000(1.08)^{9.1} = $402,899.71$; the value is more than doubled after 9.1 years.

8. This property would be worth $252,041 after 10 years, so the original property is the better investment.

9. 3.18 acres; 1.59 acres; 1.06 acres

10. As the number of lots increases, the size of each lot decreases. The relationship is reciprocal.

11. $y = \dfrac{318}{x}$

12.

13. about 254   **14.** $3,810,000

15. 0.795 acres   **16.** $3,800,000

17. The relationship is linear. The differences between successive AGI values is a constant $28,400.

18. $70,900   **19.** $51,020

20. The relationship is quadratic. The second differences are a constant $1400.

21. $71,700   **22.** decrease

23. The percentage rate is constant and is not related to the price. The dollar amount charged increases linearly with the price of the house. The slope of this increasing line is the commission percentage.

24. piecewise function

25. The homeowner would pay a lower commission to a real estate agent if his or her home is worth $100,000 or more.

26. The agent would be more likely to attract homeowners with properties over $100,000 and would make large commissions even at the reduced commission rate.

Tech Prep Applications— Chapter 11

Quality Assurance

1.

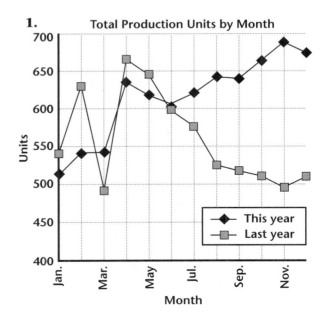

2. November **3.** 196 units

4. this year: 614.6 units; last year: 558.9 units

5. Answers may vary. Possible answer: The production has increased this year by an average of more than 50 units per month; this suggests that the new managers are running a more efficient manufacturing plant.

6. mean = 19.83; median = 18; mode = 18; range = 22

7. mean = 3.3%; median = 3%; mode = 3%

8.

Stem	Leaves
0	9
1	6 7 8 8 8 8
2	0 2 4 7
3	1

9. upper quartile = 17.5; lower quartile = 23

10.

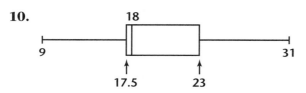

11. $\frac{7}{150}$ = .047, or 4.7%; this percent is higher than average.

12. Answers may vary. Sample answer: Inform the process engineering department that the changes were ineffective.

13. power source: 17.2%; LEDs: 24.8%; mechanical: 29.2%; interfaces: 10.8%; wiring: 6%; exterior: 12%

14.

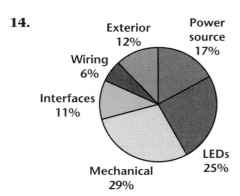

15. power source: 258; LEDs: 372; mechanical: 438; interfaces: 162; wiring: 90; exterior: 180

16.

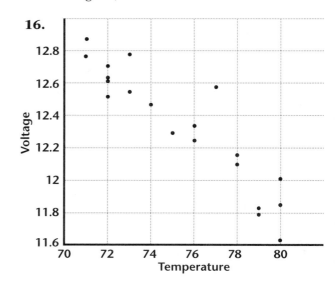

17. There is a strong negative correlation in the data.

18. Answers may vary. Sample answer: Control the temperature of the machine which installs power sources so that the temperature remains at or below 74°F.

Tech Prep Applications— Chapter 12

Aquaculture

1.

A	400 ft³	600 ft³
B	500 ft³	750 ft³
C	1200 ft³	1800 ft³
D	1500 ft³	2250 ft³
E	1750 ft³	2625 ft³

2. 3 times **3.** 3 times

4. The volume in pond C would still be 3 times larger than the volume in pond A because the depth is a common multiplier that influences both volumes equally.

5. 3 times **6.** 3.5 times

ANSWERS

7. A larger pond can mathematically equal the surface area (and volume) of three smaller ponds. However, smaller ponds would not be placed along side one another to equal a larger pond and would need more total space around them than a larger pond. Therefore, larger ponds probably occupy less land area.

8.

6 m	28.3 m²	14.1 m³
8 m	50.3 m²	25.1 m³
10 m	78.5 m²	39.3 m³
12 m	113.1 m²	56.5 m³

9. 4 times   **10.**   4 times   **11.** 10,000

12. 7500   **13.** ≈ 132.7 m²   **14.** ≈ 1.33 kg

15. 433 gal   **16.** 2684 lb   **17.** ≈ 66 ft²

18. 9.1-ft diameter circle

19. $L_{cone} + SA_{cyld} - A_{cyld\ base} = 42.4$ ft²

20. Calculate the perimeter of a 3-cm-by-3-cm square, which is 12 cm. This would be the minimum girth (distance around) of a fish that would fit through the gill net. The size or weight of the fish depends on its species.